About the book

There are 4, year 9 Mathematics papers & answers in this book. These are 2 sets of papers 1 (non-calculator) & 2 (calculator) written as practice papers for end of year 9 Mathematics Examinations in June 2020. Papers are mainly focusing on topics covered by most schools in year 9 mathematics syllabuses in The United Kingdom. However, you may still use this book as a practice for other syllabuses for 13 to 14 year olds.

All the questions in this book are written by the author and they are new questions written purely to help and experience the students to prepare and test themselves for the upcoming end of year mathematics exams.

Answers are included in this book. If you need to check your solutions, I advise you to ask your school mathematics teacher or your private mathematics tutor to mark your answers.

There are 2 sections to this book A & B. Each section contains 2 papers. The first paper of each section is a non-calculator paper & the second paper of each section is a calculator paper.

Year 9 Mathematics Practice Papers

(Year 9 Mock Exams)

for 13 to 14 year olds

4 mock papers including answers

By Dilan Wimalasena

Contents

Section A

Year 9

Mathematics

Practice Paper A1

June 2020

Calculator is not allowed

**Time allowed
1 hour
Total 100 marks**

Write answers in the space provided

1. Write the following numbers in standard form.
i) 2351

(2 marks)

ii) 23.51

(2 marks)
(total 4 marks)

2. Convert the following into centimetres.
i) 12m

(2 marks)

ii) 1.24m

(2 marks)

iii) 2km

(4 marks)
(total 8 marks)

3. Expand and simplify.
$i)\ (x + 3)(x + 5)$

(3 marks)

$ii)\ (x - 3)(x - 3)$

(3 marks)

$iii)\ (2x + 3)(x - 5)$

(3 marks)

$iv)\ (4x - 1)(3x - 2)$

(3marks)
(total 12 marks)

4. Work out an exterior angle of a regular pentagon.

(4 marks)

5. John rolled a six-sided dice.
Work out the probability of getting a multiple of 3.

(3 marks)

6. Work out the following.

i) $\dfrac{2^3 \times 2^8}{2^5 \times 2^2} =$

(3 marks)

ii) $(3^5)^3 \times (3^8)^{-2} =$

(4 marks)
(total 7 marks)

7. Work out

$$3\frac{1}{2} \div \left(4\frac{1}{3} - 2\frac{3}{4}\right)$$

(6 marks)

8. Solve the following inequalities.

$i)\ 3x - 1 > 14$

(3 marks)

$ii)\ 4x - 2 \leq 16 - 5x$

(4 marks)

(total 7 marks)

9. The areas of the triangle and the rectangle are equal. Find the value of x.

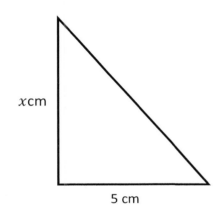

x cm

5 cm

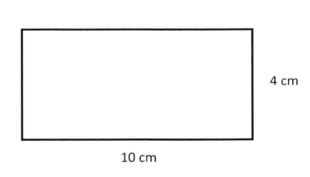

4 cm

10 cm

(4 marks)

10. Information for Dogs' age and weight in a veterinary surgery is given below.

Age (months)	1	2	3	4	5	6
Weight (kg)	2.2	2.3	2.8	3.4	6	10

i) Represent the data on a scatter graph on the grid below.

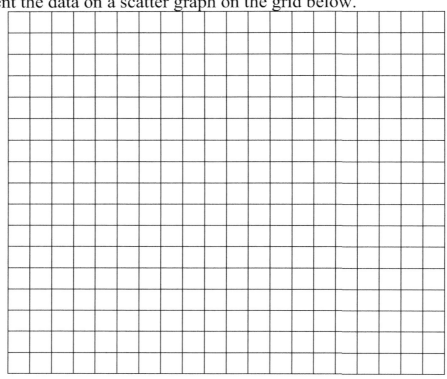

(3 marks)

ii) What type of a correlation does the scatter graph suggest?

(1 marks)
(total 4 marks)

11. Write 1200 as a product of its prime factors.

(4 marks)

12. Andrew bought a car for £12,000 and sold it for £15,000.
Work out his percentage profit.

(3 marks)

13. A liquid has density of $3g/cm^3$. Work out the mass of $2\ l$ of this liquid.

$$Density = \frac{mass}{volume}, \qquad 1cm^3 = 1ml$$

(4 marks)

14. Solve the following equations.
$i)\ 5(x-2) - 2(2x+1) = 3$

(3 marks)

$ii)\ x + y = 10$
$\quad\ x - y = 2$

(5 marks)
(total 8 marks)

15. Factorise fully.
i) $x^2 + 7x + 12$

(3 marks)

ii) $x^2 - 7x - 30$

(3 marks)

iii) $x^2 + 3x - 4$

(3 marks)

iv) $x^2 - 8x + 15$

(3 marks)
(total 12 marks)

16. John sells 5kg bags of rice for £8 and Brian sells 3kg bags of rice for £7.50. Who is better value? (show clear working)

(5 marks)

17. Construct a triangle whose sides have lengths 8cm, 6cm & 5cm.

(5 marks)

Total for paper: 100 marks

End

Year 9

Mathematics

Practice Paper A2

June 2020

Calculator is allowed

Time allowed
1 hour
Total 100 marks

Write answers in the space provided

1. John, Mary & Claire shared £240 in ratio 1:2:9. How much did Claire receive more than Mary?

(4 marks)

2. The line l_1 has equation $y = 3x - 1$. Write down the equation of line l_2, which is perpendicular to l_1 and passes through (0,7).

(2 marks)

3. Work out the following and write your answers in standard form.

i) $(2.4 \times 10^8) \times (1.2 \times 10^{-5})$

(3 marks)

ii) $\dfrac{3.5 \times 10^{10}}{7 \times 10^{-2}}$

(3 marks)
(total 6 marks)

4. David drove 120km in 1.5 hours.
i) Calculate his average speed in km/h.

(3 marks)

ii) Work out his average speed in m/s.

(3 marks)
(total 6 marks)

5. Work out the areas of following shapes.
i) A circle with a radius of 7.5cm.

(2 marks)

ii) A square with a length of 8.3cm.

(2 marks)

iii) A triangle with a height of 10.2cm & a base of 5.6cm.

(3 marks)
(total 7 marks)

6. A table is £150. Calculate the price of the table after a 17.5% discount.

(3 marks)

7. The chance that Andrew will be late to school is 0.1 and the chance that Barry will be late to school is 0.15.

i) Represent the above information on a tree diagram.

(4 marks)

ii) Calculate the probability of both of them will be on time.

(4 marks)
(total 8 marks)

8. Work out the following.

i) $2^3 \times 3^2$

(3 marks)

ii) $\dfrac{(4^3)^2 \times (2^3)^5}{(8^2)^3}$

(5 marks)
(total 8 marks)

9. Work out highest common factor (HCF) & lowest common multiple (LCM) of following.

i) 96 &108

(4 marks)

ii) 120 & 150

(4 marks)
(total 8 marks)

10. A is directly proportional to B. When A = 120, B = 24.
Work out the value of A, when B = 36.

(4 marks)

11. i) Convert $250cm^2$ to mm^2.

(3 marks)

ii) Convert $2km^3$ to m^3.

(3 marks)
(total 6 marks)

12. Work out the value x in each triangle.

i)

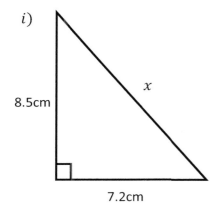

ii)

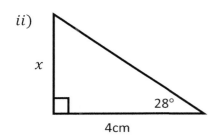

(3 marks)

(3 marks)

iii)

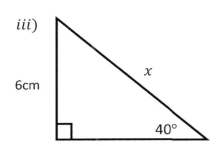

(3 marks)
(total 9 marks)

13. The volume of the cuboid below is $120cm^3$.

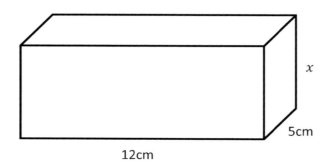

i) Work out the value x.

(3 marks)

ii) Hence, calculate the surface area of cuboid.

(3 marks)
(total 6 marks)

14. A sequence has terms 2.7, 3.3, 3.9, ….…...
i) Find an expression for the nth term of the sequence.

(2 marks)

ii) Find the 100th term of the sequence.

(2 marks)

iii) Is 20.1 a term in the sequence.

(3 marks)
(total 7 marks)

15. Weight of a car is 1250kg correct to 3 significant figures.
Write down the upper and the lower bounds for the weight of the car.

(4 marks)

16. $A(2,5), B(3,8)$

i) Calculate the gradient of the line AB.

(2 marks)

ii) Hence, Find the equation of line AB in the form $y = mx + c$.

(4 marks)
(total 6 marks)

17. $y = x^2 - 4x + 3$

i) Compete the table below.

x	-1	0	1	2	3	4	5
y							

(3 marks)

ii) Plot the graph of $y = x^2 - 4x + 3$ on the grid below.
(clearly label your x & y axes)

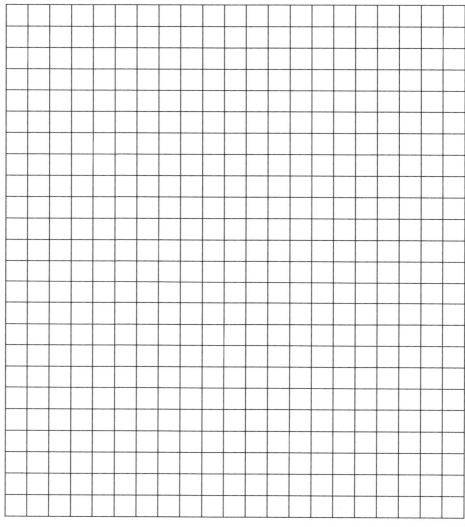

(3 marks)
(total 6 marks)

Total for paper: 100 marks

End

Section B

Year 9

Mathematics

Practice Paper B1

June 2020

Calculator is not allowed

Time allowed
1 hour
Total 100 marks

Write answers in the space provided

1. Simplify fully.

i) $(x^2)^3$

(1 mark)

ii) $\dfrac{x^2 \times x^5}{x^3}$

(2 marks)

iii) $\dfrac{(x^2)^3 \times x^5}{x^7 \times (x^3)^2}$

(3 marks)

(total 6 marks)

2. Write the following as percentages.

i) 0.35

(2 marks)

ii) $\dfrac{3}{5}$

(3 marks)

iii) $\dfrac{4}{25}$

(3 marks)

(total 8 marks)

3. John, Sam & Freddy shared £360 in the ratio 2:6:7. Work out each share?

(4 marks)

4. Make x the subject of the following formulae.

i) $y = 2x - 3$

(2 marks)

ii) $a = bx^2 - c$

(3 marks)

iii) $y = \dfrac{3x + 4}{5}$

(3 marks)

(total 8 marks)

5. Prove that the triangle ABD is congruent to the triangle ACD.

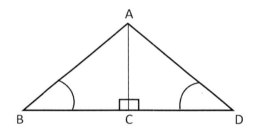

(4 marks)

6. Estimate the following.

$$\frac{(2.49 \times 3.99)^2}{\sqrt{99.8}}$$

(3 marks)

7. The table below shows the weights of a group of 20 people.

Weights (kg)	Frequency (f)
40-60	3
60-70	5
70-80	7
80-90	4
90-100	1

i) Write down the modal class.

(2 marks)

ii) Work out the median class.

(3 marks)

(total 5 marks)

8. Write 150 as a product of its prime factors.

(3 marks)

9. Work out

$$3\frac{1}{3} - \left(4\frac{1}{2} \div 1\frac{3}{4}\right)$$

(5 marks)

10. Expand and simplify.

i) $(2x - 3)(x + 5)$

(3 marks)

ii) $(3x + 1)^2$

(3 marks)
(total 6 marks)

11. Solve the following equations.

i) $\dfrac{3x}{5} - 5 = 7$

(3 marks)

ii) $2(3x + 1) + 4(2x - 9) = 10$

(4 marks)
(total 7 marks)

12. Factorise fully.

i) $24x - 36x^2$

(2 marks)

ii) $x^2 - 9$

(3 marks)

iii) $x^2 - 13x - 30$

(3 marks)
(total 8 marks)

13. Work out the angles x & y. Given that lines BD & EF are parallel. (write reasons for each stage of your working)

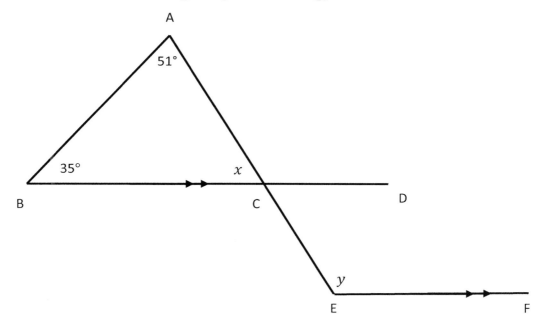

(5 marks)

14. Work out

$$2.34 \times 1.98$$

(4 marks)

15. Work out the volume of cylinder in terms of π.

Radius = 3cm

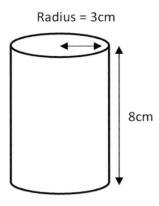

8cm

(4 marks)

16. Write the following numbers in standard form.

i) 234

(2 marks)

ii) 23.4

(2 marks)

iii) 0.234

(2 marks)

iv) 0.00234

(2 marks)
(total 8 marks)

17. Work out 48% of £350.

(4 marks)

18. Convert the following into m^2.

i) $24000cm^2$

(4 marks)

ii) $2km^2$

(4 marks)
(total 8 marks)

Total for paper: 100 marks

End

Year 9

Mathematics

Practice Paper B2

June 2020

Calculator is allowed

Time allowed
1 hour
Total 100 marks

Write answers in the space provided

1. Work out highest common factor (HCF) and lowest common multiple (LCM) of

i) 68 & 102

(4 marks)

ii) 120 & 180

(4 marks)

(total 8 marks)

2. A rectangle has length 24.5cm correct to 1 decimal place and width 10cm correct to 1 significant figure.

Calculate the upper bound for area of the rectangle.

(4 marks)

3. John invested £2500 in an account for 2 years. Bank offers 4% compound interest per annum.

Work out his balance after 2 years.

(4 marks)

4. $y = x^2 + 2x - 5$
i) Complete the table

x	-4	-3	-2	-1	0	1	2
y							

(3 marks)

ii) Plot $y = x^2 + 2x - 5$ on the grid below.

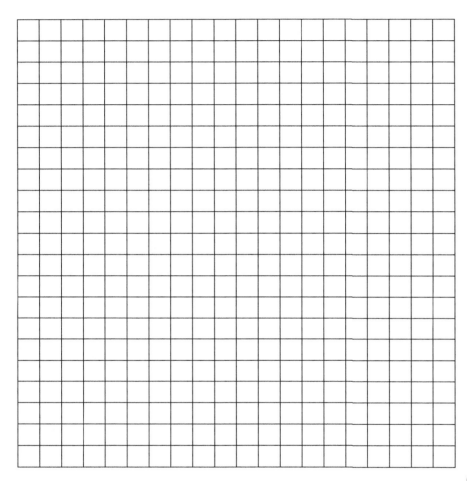

(3 marks)
(total 6 marks)

5. Work out the sum of interior angles of an octagon.

(3 marks)

6. Solve the following equations.

i) $\dfrac{2x - 5}{3} = 4x + 1$

(3 marks)

ii) $\dfrac{2x + 3}{5x - 1} = \dfrac{2}{7}$

(4 marks)

(total 7 marks)

7. OCD & OAB are triangles as shown below. CD & AB are parallel.

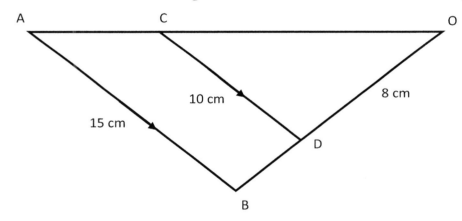

i) Prove that triangles OCD and OAB are similar.

(3 marks)

ii) Hence, work out the length BD.

(4 marks)

(total 7 marks)

8. Work out the value of x to 1 decimal place in each triangle.

i)

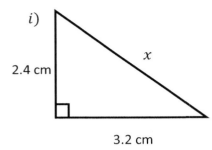

2.4 cm

x

3.2 cm

(3 marks)

ii)

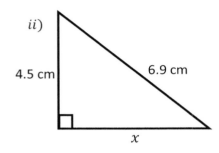

4.5 cm

6.9 cm

x

(3 marks)

iii)

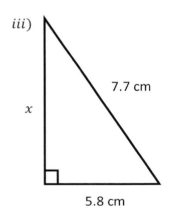

7.7 cm

x

5.8 cm

(3 marks)
(total 9 marks)

9. Wok out the following and write your answer in standard form.

$$(2.8 \times 10^2) \times (4 \times 10^{-3})$$

(3 marks)

10. Solve the following simultaneous equations.

$i) \ a - 2b = 3$

$\quad 5a - 2b = 31$

(4 marks)

$ii) \ 2x + 3y = 17$

$\quad 5x - 3y = 11$

(4 marks)

(total 8 marks)

11. P is inversely proportional to Q^2. When $Q = 2, P = 15$. Find P, when $Q = 10$.

(5 marks)

12. $i)$ Convert 20km/h into m/s.

(3 marks)

$ii)$ Convert 4m/s into km/h.

(3 marks)

(total 6 marks)

13. Cathy bought 6kg of pasta for £1.86.
Work out the cost of 14kg of pasta.

(4 marks)

14. Factorise fully.
i) $x^2 - 17x + 60$

(3 marks)

ii) $x^2 - 17x - 60$

(3 marks)

iii) $x^2 + 17x - 60$

(3 marks)
(total 9 marks)

15. A circle has circumference 29cm.
i) Calculate the radius of the circle.

(3 marks)

ii) Calculate the area of the circle.

(2 marks)
(total 5 marks)

16. Diagram below shows the triangle PQR. The side PS is perpendicular to side QR.

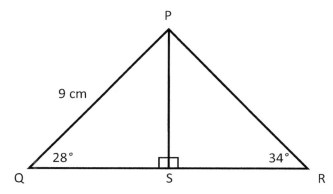

Calculate

i) PS

(3 marks)

ii) QS

(3 marks)

iii) RS

(3 marks)

iv) Area of triangle PQR

(3 marks)
(total 12 marks)

Total for paper: 100 marks

End

Answers

Paper A1	Paper A2
1.i) 2.351×10^3, ii) 2.351×10^1	1. £140
2. i) 1200cm, ii) 124cm, iii) 200,000cm	2. $y = -\frac{1}{3}x + 7$
3. i) $x^2 + 8x + 15$, ii) $x^2 - 6x + 9$, iii) $2x^2 - 7x - 15$, iv) $12x^2 - 11x - 2$	3. i) 2.88×10^3, ii) 5×10^{11}
4. $108°$	4. i) $80km/h$, ii) $22.2m/s$
5. $\frac{1}{3}$	5. i) $176.71cm^2$, ii) $68.89cm^2$, iii) $28.56cm^2$
6. i) 16, ii) $\frac{1}{3}$	6. £123.75
7. $2\frac{4}{19}$	7. i) correct tree diagram, ii) 0.765
8. i) $x > 5$, ii) $x \leq 2$	8. i) 72, ii) 4096
9. 16cm	9. i) $HCF = 12, LCM = 864$, ii) $HCF = 30, LCM = 600$
10. i) correct plot, ii) positive correlation	10. 180
11. $2 \times 2 \times 2 \times 2 \times 3 \times 5 \times 5$	11. i) $25,000mm^2$, ii) $2,000,000,000m^3$
12. 25%	12. i) $11.14cm$, ii) $2.13cm$, iii) $9.33cm$
13. 6000g	13. i) $2cm$, ii) $188cm^2$
14. $x = 15$	14. i) $0.6n + 2.1$, ii) 62.1, iii) Yes with reasons
15. i) $(x + 3)(x + 4)$, ii) $(x - 10)(x + 3)$, iii) $(x + 4)(x - 1)$, iv) $(x - 3)(x - 5)$	15. upper bound 1255kg, lower bound 1245kg
16. John with reasons.	16. i) $m = 3$, ii) $y = 3x - 1$
17. correct construction using compass & ruler.	17. i)

17. i)

x	-1	0	1	2	3	4	5
y	8	3	0	-1	0	3	8

ii) correctly plotted curve.

Paper B1	**Paper B2**
1. i) x^6, ii) x^4, iii) x^{-2}	1. i) $HCF = 34, LCM = 204$, ii) $HCF = 60, LCM = 360$
2. i) 35%, ii) 60%, iii) 16%	2. $368.25cm^2$
3. John £48, Sam £144, Freddy £168.	3. £2704
4. i) $\frac{y+3}{2} = x$, ii) $\sqrt{\frac{a+c}{b}} = x$, iii) $\frac{5y-4}{3} = x$	4. i)
5. $AC = common\ side$ $\angle ABC = \angle ADC\ (given)$ $\angle ACB = \angle ACD\ (right\ angle)$ $\therefore \Delta ABC \equiv \Delta ADC\ (AAS)$	ii) correctly plotted curve
6. 10	5. $1080°$
7. i) 70-80kg, ii) 70-80kg	6. i) $x = -\frac{4}{5}$, ii) $x = -\frac{23}{2}$
8. $2 \times 3 \times 5 \times 5$	7. i) $\angle COD = AOB\ (common)$ $\angle OCD = \angle OAB\ (corresponding\ angles)$ $\angle ODC = \angle OBA\ (corresponding\ angles)$ $\therefore \Delta OCD$ is similar to ΔOAB. ii) 4cm
9. $\frac{16}{25}$	8. i) 4cm, ii) 5.23cm, iii) 5.06cm
10. i) $2x^2 + 7x - 15$, ii) $9x^2 + 6x + 1$	9. 1.12×10^0
11. i) $x = 20$, ii) $x = 3\frac{1}{7}$	10. i) $a = 7, b = 2$, ii) $x = 4, y = 3$
12. i) $12x(2 - 3x)$, ii) $(x + 3)(x - 3)$, iii) $(x - 15)(x + 2)$	11. $P = 0.6$
13. $x = 94°\ (angles\ in\ a\ triangle)$, $y = 86°\ (allied\ angles)$	12. i) $5.56m/s$, ii) $14.4km/h$
14. 4.6332	13. £4.34
15. $72\pi cm^3$	14. i) $(x - 12)(x - 5)$, ii) $(x - 20)(x + 3)$, iii) $(x + 20)(x - 3)$
16. i) 2.34×10^2, ii) 2.34×10^1, iii) 2.34×10^{-1}, iv) 2.34×10^{-3}	15. i) $r = 4.62cm$, ii) $A = 67.06cm^2$
17. £168	16. i) $PS = 4.23cm$, ii) $QS = 7.95cm$, iii) $RS = 6.27cm$, iv) $30.08cm^2$
18. i) $2.4m^2$, ii) $2,000,000m^2$	

Paper B2, Question 4. i)

x	-4	-3	-2	-1	0	1	2
y	3	-2	-5	-6	-5	-2	3

Printed in Great Britain
by Amazon